DOING
PRIMARY
RESEARCH

VALERIE BODDEN | ILLUSTRATIONS BY ELWOOD H. SMITH

C R E A T I V E E D U C A T I O N

Published by Creative Education
P.O. Box 227, Mankato, Minnesota 56002
Creative Education is an imprint of The Creative Company
www.thecreativecompany.us

Design and production by Liddy Walseth
Art direction by Rita Marshall
Printed in the United States of America

Illustrations by Elwood H. Smith © 2012

Library of Congress Cataloging-in-Publication Data

Bodden, Valerie.
Doing primary research / by Valerie Bodden.
p. cm. — (Research for writing)
Includes bibliographical references and index.
Summary: A narrative guide to conducting primary research, complete with
an overview of methodologies, tips for collecting and applying
qualitative and quantitative data, and helpful resources.
ISBN 978-1-60818-204-6
1. Research—Methodology—Juvenile literature. 2. Information literacy—Juvenile literature. I. Title.
ZA3080.B63 2012
001.4'2—dc23   2011040490

First Edition
2 4 6 8 9 7 5 3 1

TABLE OF CONTENTS

YOU CONDUCT
RESEARCH EVERY DAY.
AND YOU PROBABLY DON'T EVEN REALIZE IT.

When you want to know the score of last night's football game, you check on-line. If you're wondering how to spell (or define) "supercilious," you get out the dictionary. And when you've forgotten the math assignment, you e-mail your friends to ask them. Although these research situations are rather informal, the skills you use for them are similar to those you need to conduct research for writing. After all, research is basically just a search—a search for sources that contain specific, **relevant**, and accurate information about your topic.

But where do you look for that information? Well, that depends on what you want to know. For some subjects, books will be your best sources. For others, you might find what you need on the Internet. In many cases, you'll need to do primary research—that is, research that contains firsthand information. Let's say your assignment is to write about the 2010 Winter Olympic Games, for example. Today's newspaper probably isn't going to cover the event. You will most likely find an entry in an encyclopedia, but will that be enough? How about turning right to the source—including television footage of the Games and interviews with the athletes?

Primary sources are a great way to learn about an event from the perspective of those who experienced it. And you can conduct your own primary research, too, by carrying out experiments, surveys, interviews, and observations. Such firsthand information can back up old claims and even lead to new insights.

PRIMARY SOURCE PRIMER

MOST OFTEN, YOU WILL HEAR PRIMARY SOURCES referred to in conjunction with secondary sources. But why are they called primary and secondary sources? Is it because primary sources are more important than secondary sources? Actually, no. Primary sources are called "primary" because of their direct, firsthand nature. Nothing stands between the original source and the event or information it relates. In contrast, secondary sources report on—and often interpret—primary sources. They rely on the information-gathering and analytical skills of a secondary author—someone who was not present for an event or was uninvolved in the original experimental or surveying process. For example, if you are writing a paper about the works of J. K. Rowling, *Harry Potter and the Sorcerer's Stone* would be a primary source. An article offering literary criticism or analysis of the novel would be a secondary source.

Both primary and secondary sources can be valuable tools when doing research for a writing assignment. In fact, you'll often want to turn to secondary sources first. Why? Secondary sources can provide needed background information on a subject—and unless you have that background information, a primary source might not make much sense to you. For example, if you are researching Abraham Lincoln's Gettysburg Address but do not know where or when or why he gave the speech, it may be difficult to understand. Secondary research can also help you prepare for conducting your own primary research, such as an interview. It will be hard to know what to ask an expert about the abolition of slavery, for example, if you don't know something about the subject in the first place.

That said, primary sources can be an excellent way to support or prove a point. So when should you use primary research—and what kinds of primary research should you use? Often, that depends on the subject. If you are researching a historical topic, for example, you might turn to original documents, paintings, maps, or historical **artifacts**. Newspaper articles or radio or television broadcasts made at the time of an event can also serve as primary sources. Video footage of the September 11, 2001, terrorist attacks on the World Trade Center in New York could be considered a primary source for research on that subject, for example. A paper about literature will most likely focus on a specific novel or story—in this case, the text is the primary source. If your assignment is to research a particular person, primary sources may include that person's diary, letters, or **autobiography**. For scientific subjects, articles detailing the results of new research could be primary sources, as they often incorporate summaries or charts of data obtained through experiments or observations. Many of these types of primary sources can be found at the library, in an **archive**, or even on the Internet.

NOT ALL PRIMARY RESEARCH CAN BE FOUND

Not all primary research can be found, however. Sometimes it has to be created. And you are the one who gets to do the creating! If you are interested in subjects on which little previous research has been done, subjects based on issues specific to your school or town, or subjects on which you want an insider's view, you'll want to perform your own primary research. How you go about doing this will depend on the topic. If it is scientific, you might set up your own experiment. Or, if it has to do with the behavior of people (or animals), you could consider making observations, in which you carefully watch the actions of a specific group or event. To obtain specific information from a certain group of people in your community, you might create a survey. For example, maybe you want to know whether people support a new law. Or you might want to verify that the results of another researcher's work hold true in your community. While a national survey might reveal that 31 percent of respondents do not exercise at all, you may think the people in your town are more (or less) health conscious. You can create your own survey to find out. In some cases, you might want an eyewitness account or an expert interpretation of an event, so you would hold an interview.

The keys to conducting good primary research are knowing what you are looking for, where to find it, and how to evaluate it. These skills are known as information literacy, and according to the American Library Association, a person who has mastered them has "learned how to learn." The first step in the process of becoming information literate—and in doing any research, primary or otherwise—is to determine what kind of information you need to find. That might seem obvious, but it takes more thought than it might at first appear. For example, if your assignment is to write a paper about some as-

LEARNED HOW TO LEARN

pect of television, you are quickly going to hit a wall if you simply start looking for sources on "television." The topic is too broad—you will end up with results about the history of television, how TV sets work, products to buy, specific shows and channels, and more. So, before you begin your search, you must decide what it is about television that you want to study. Perhaps, after you do some thinking (and maybe some reading in a secondary source to help narrow your topic), you will decide that you want to focus on how advertising influences young TV viewers.

Once you have narrowed your topic, the next step in becoming information literate is to locate information related to it. There are secondary sources—from books to encyclopedias to journal articles—available on nearly every topic imaginable. But think about what primary sources you might use for your research as well. In the case of the paper on advertising and young TV viewers, for example, studies about the subject may have been published

in scholarly journals. You could also conduct your own primary research into the topic by observing, surveying, or interviewing young people about their television viewing habits and the ads they see.

By the time you have collected your information, you may think that your work is done. It's not. (Surprise!) An important step in the quest to becoming information literate is analyzing and evaluating your information. If you are relying on primary research that you did not carry out yourself, consider

the source. Is the information current? Who designed the experiment or performed the investigation? Is he a recognized expert in the field? Does the information appear to be accurate? You can determine this by comparing it with other sources. If you are studying primary documents such as autobiographies or letters, you also need to consider whether the author is **biased**. Many people might try to avoid or put a positive spin on certain aspects of their life when they are recording it for others to read.

Don't be afraid to evaluate the results of your own research as well. Does your survey or experiment agree with the results of other researchers? Could your questions have been **leading** or biased, resulting in skewed results? Was the person you interviewed a true authority on the subject?

In addition to considering the accuracy of your primary sources, think about their relevance. Do they provide you with information you need to know? No matter how accurate the information is, if it doesn't address your topic, it won't do you any good.

Finally, once you've found and evaluated your sources, be sure to use them correctly. Keep track of each source's author, title, publisher, and publication date, and if you use an idea that isn't your own, be sure to **cite** where it came from. Likewise, if you use someone else's words directly, put them in quotation marks. So pick through your information to find the most accurate, relevant sources available—the result will be a strong foundation on which to build your writing assignment.

EYE ON EVALUATION

You can practice evaluating information for its accuracy by comparing two works about the same subject. First, find an autobiography by a prominent individual who interests you. Next, find a biography about that same person. (An autobiography is written by the subject herself, a biography by someone else.) Look for sections of each book that describe the same event, using the table of contents or index to find references to the incident in each. Do both books give similar details? Make a list of any differences you spot. Do the differences reveal any bias on the part of either author (perhaps a desire by the subject to portray her actions in a more positive light, for example)? Although the autobiography will still likely be a strong primary source—because it is in the words of your subject—being aware of such biases can help you know how to interpret what she writes.

METHODS
TO COUNT ON

WHEN YOU DO YOUR OWN PRIMARY RESEARCH, you first need to decide whether you want to perform quantitative or qualitative research. Quantitative research focuses on numbers and measurements that can be recorded and analyzed to draw conclusions. The results of qualitative research, on the other hand, are typically expressed in words. So when you want to answer questions of "how much," "who," or "how many," you will likely rely on quantitative research such as experiments or surveys. A benefit of quantitative data is that it can be used to prove or disprove a **hypothesis**. However, it usually cannot answer questions of "why." Quantitative data often requires the use of specific data collection tools, such as measuring devices or survey forms.

If, instead of answering questions of "how much," you want to know "why" or "how," you will probably turn to qualitative research in the form of interviews and observations. Qualitative research can help you understand a particular issue and its effect on people. The only data collection tool is typically you, the researcher. This can be an advantage, since you don't have to worry about using a poorly designed tool. But it can also be a disadvantage, as the results hinge on your own interpretation of what you see, which might lead to inconsistencies or misrepresentations.

HOW MUCH HOW MANY

A careful examination of your topic should lead you to a specific research method (or perhaps a combination of methods) that will give you some answers. Let's say that you decide your project calls for quantitative research gathered through an experiment. Now how do you go about setting up that experiment? First, determine exactly what it is you want **GIVE YOU** to test or find. Are you looking for the ideal wind speed for flying a kite? Then your research will need to take place in the **field**. Do you want to know how much liquid different materials can absorb? For that, you'll want to work in a lab (or in your home), where you have more control over conditions.

In general, the more control you have over the conditions of your **SOME** experiment, the more accurate your results will be. For example, if you are testing the absorbency of materials, you'll want to use the same liquid for each and soak each up off the same surface (such as a table). Outdoors, of course, you will not have as much control, but you should try to **ANSWERS** minimize other **variables** that could affect your results. In the kite experiment, for example, you would want to use the same kite, the same tool for measuring wind speed, and the same kite flier (some people have more kite flying skill than others, which could affect the results) every time you do the experiment.

If you want to know about something that cannot be easily tested or observed (people's attitudes or opinions, for example), you might consider creating a survey, in which you ask the same set of questions of several people and then tally and compare their answers. Before you design a survey, you should read the available resources on your topic and decide exactly what it is that you want to discover with your questions. Obviously, you can't survey everyone in the world or in your town—or probably even in your school. But you can survey a sample population, or a portion of a specific group. First, you need to decide who that group will be. Do you want to know how teenage girls feel about dieting? Then it won't do you any good to survey boys of any age—or to survey girls under 12 or over 20. What do you need to know about your sample population in order to make your survey more effective? Their age and gender, of course. But what about their school (will it make a difference if they go to a public or private school or are homeschooled?), siblings (does it matter if they have brothers or sisters?), parents (is divorce a factor?), and other considerations? Think about any factors that might help you interpret or further classify your results, and be sure to collect this data at the beginning of your survey.

In addition to specific information about your survey participants, you need to gather information on their opinions and behaviors. You might write yes or no questions to get direct answers: "Have you dieted in the last six months?" for example. Or you could use multiple-choice questions, asking, "How many days a week do you exercise?" and following that with

several standard choices: "A. 7, B. 5–6, C. 3–4, D. 1–2, E. 0." As you write your questions, be sure that they are clear and that any unfamiliar or vague terms are defined. In the question above, for example, you might need to define the term "exercise" so that survey participants know if you mean "walk from the refrigerator to the TV" or "run a marathon."

When dealing with **controversial** subjects, you also need to be careful not to introduce **loaded** or leading words into your questions. If, for example, you think that wearing real fur is wrong, you need to avoid revealing that through the wording of your question: "Do you agree that killing cute, innocent animals simply for the sake of looking 'cool' is awful?" would be a loaded question, for instance. You can also include open-ended questions in your survey, allowing participants to write their own short answers, in order to collect qualitative data. If your survey is about smoking habits among young adults, you might ask why, if the participant smokes, she began smoking or how she feels about smoking bans in public buildings. Although the answers to such questions will likely be harder to tabulate—since there will be more variety—they may also provide useful insights into your survey results.

When you have your survey questions ready, you can survey away! You've already identified whom you want to ask and what you want to ask them. Now, how will you do it? You can hand people a written copy of the survey and ask them to fill it out. Surveys can also be administered over the phone, online, or via e-mail. In general, the more people you survey, the more meaningful your results will be, as you will likely reach a more **representative** sample of your group. Surveying too small of a sample population may lead to skewed results as you **extrapolate** your findings onto a larger population.

20 TO 30 30 TO 40

For example, if you survey three junior high boys and learn that all three prefer playing basketball over football, you may be led to the conclusion that all junior high boys prefer basketball. If, however, you ask 30 junior high boys the same question, you are likely to get a mix of responses that probably more closely represents the opinion of the age group as a whole.

Once your survey is complete, it is time to tabulate and analyze the results. In some cases, you will simply need to figure out what percentage of people gave each answer. In other cases, you might want to take a closer look, analyzing whether there might be a connection between factors such as age, gender, or **ethnicity** and specific answers. For example, do more people between the ages of 30 and 40 report reading for pleasure than do people between the ages of 20 and 30? Open-ended questions are more difficult to tabulate, but you may be able to divide them into broad categories to determine whether any patterns emerge in participants' answers. And in the end, that is what you are looking for—patterns, numbers, and quantifiable data. So think about what numbers you need to know—and then design a way to find them!

THE SCIENTIFIC METHOD

Often, quantitative data is used to answer questions in scientific fields, such as biology, astronomy, or geology. In constructing experiments to collect such data, scientists generally follow a carefully controlled series of steps known as the scientific method. The first step is to ask a research question—the question you will answer through your experiment. For example, you might want to know, "How accurately does a cricket's chirp reflect the temperature?" Next, you will want to read available information on the subject to see what other researchers have found. Based on your reading, form a hypothesis, or an educated guess, to answer your research question. Then conduct your experiment to find out if your hypothesis is correct. Be sure to make careful observations and to record your results. Was your hypothesis correct? Even if it wasn't, you've learned a valuable lesson in how to collect quantitative data!

WONDERING WHY

ALTHOUGH QUANTITATIVE RESEARCH IS USEFUL in many situations, sometimes there is no way to put a number to a topic. Or there may be a way to quantify the data, but you want to get beyond numbers to learn more about how something occurs or why someone feels a certain way. If, for example, you are researching cancer, you could survey cancer patients to find out what kinds of cancer they suffer from, whether they feel depressed, or what kinds of treatment methods they are trying. But what if you want to know *how* their cancer affects their daily lives, *why* they feel depressed, or *what* led them to make their treatment choices? In that case, you need to turn to qualitative research methods such as interviews.

Whom you choose to interview depends on both your topic and what you want to learn about it. If you are looking for **anecdotes**, eyewitness accounts, or descriptions, you might talk to someone who experienced an event firsthand, even if he is not an expert in the subject. A World War II bomber pilot, for example, may not know the date of every major battle in the war, but he could tell you fascinating stories about his time in a prisoner-of-war camp or flying in bombing raids on enemy cities. On the other hand, if you are looking for an interpretation of an event's significance or an explanation of how something works, you will most likely turn to an expert on the topic. But how do you find an expert? Local universities

have professors who are experts in a number of fields. Politicians and lawyers can give expert interpretations of local laws, and even local businesses can provide experts in their areas. You can also try to locate the author of an article or book who might be willing to participate in a phone or e-mail interview if he does not live nearby.

You shouldn't go into an interview expecting your interviewee to tell you everything there is to know about your topic. That's what secondary sources are for. You should consult such sources *before* going to your interview. Having a working knowledge of the topic will allow you to ask more specific questions, directed at obtaining exactly the information you need. For example,

YES

if you are researching civil rights protests of the 1960s, you will first want to read sources that reveal what the protests involved, who generally took part in them, and where they were centered. Then, you might interview someone who participated in the protests, asking her about her particular experiences—what, where, and, perhaps most importantly, why she did what she did. Even if you are interviewing an expert, you should expect that person to add insights to your topic and not to serve as an encyclopedic reference on it. Also, think about what your interview subject is likely to know, and plan appropriate questions. You probably wouldn't ask a circus clown how to walk on the tightrope, for example, but you might ask him the best way to get a crowd laughing.

Before arriving for your interview, write down several specific questions. Try to avoid asking anything that could be answered with a simple yes or no. Instead, ask open-ended questions intended to draw out the interviewee and learn about his experiences, opinions, or ideas. You might ask for stories about an event or an explanation of a hard-to-grasp aspect of your subject. When you actually begin interviewing the person, you do not have to follow this list as a script. Your questions are simply a guide to help you gather the information you need—and a way to get the conversation back on track if your interviewee wanders too far off topic.

NO

But if your interviewee adds an interesting point that you would like to follow up on, do it! Be sure you have some way to record what the person says (other than your memory, which can be faulty), whether it be a pencil and paper or an audio or video recorder—but ask the interviewee's permission before using such equipment. If the interview is held in your subject's home or office, you may also want to take note of the setting. Is your scuba diving expert's office filled with sunken treasures or books on the subject, for example? Such details can add color to the report of your eventual findings.

Even if you cannot meet with an interview subject in person, you might still be able to get your questions answered over the phone or via e-mail. The e-mail addresses of many experts are relatively easy to find through university Web sites or online directories. Before you e-mail a list of questions to someone you've never met, however, it is polite to first send a message requesting an e-mail interview. If the person agrees, then send her your questions—but try to limit them to four or five, since your expert is likely busy and will have to take the time to sit down and write her answers. If you need to be able to immediately ask follow-up questions, you might use an instant messaging or chat service for your interview instead.

In some cases, you might want to ask a group of people the same set of open-ended questions. You could interview each person individually, or you could hold a focus group. In a focus group, you gather a small group of people who share some characteristic, such as age or gender. Then you ask them questions, allowing each to answer—and to play off one another's answers. Advertisers use this technique, for example, if they want to understand what attracts a specific group to a new product.

If you want to get a firsthand glimpse of people's behavior, then you will rely on another type of qualitative research—

FIRSTHAND GLIMPSE

observations. Although observing something might seem simple enough, it takes planning and practice. First, as with any type of primary research, you want to read up on your topic. If you are researching the attention span of young students, for example, you will first want to read any other studies that have been done on the subject. Then, figure out what it is that you want to observe—and where you might find it. In the case of kids' attention spans, you might visit an elementary school classroom to watch how long children

focus on a single activity. (Be sure to get the school's permission first.) Once you know what you want to observe, you need to decide how you are going to do so. Will you sit to the side and just watch? Will you interact with the people you are observing? Will you take notes or use a video camera? Any of these decisions might affect your results. If people know they are being observed, they may take special care to be on their best behavior.

GIVE GREATER DEPTH TO RESEARCH

However you decide to observe, the key to actually carrying out your observation is to pay attention. But to what? You can't possibly note everything about a scene at any given moment. Instead, look for those things that answer your research question. In the case of children's attention spans, for example, you might decide to observe what kinds of activities kids focus on, how much time they devote to each, or what types of distractions call students' attention away from assigned tasks. You might also consider collecting some quantitative data on the subject by counting the number of instances students' attention seems to wander within a specific time period. This combination of qualitative and quantitative data will give greater depth to your research. When you're done observing, you'll be able to put what you see—or hear—into words that can tell others more about your topic!

ASK AWAY

The best way to get good at interviewing is to practice. So, choose a prominent person from your community about whom you would like to know more (a teacher, the mayor, a police officer, etc.) and ask if you can interview him or her. Before the interview, try to find any background information that might have been written about the person (maybe an article in the local newspaper's archives, for example). Then, write a list of questions. Perhaps you want to ask a teacher why he chose to teach a certain subject, or maybe you want to question the mayor or a police officer about a new law (be sure to read up on the topic first, too). When you go to your interview, listen attentively, take notes, and don't be afraid to ask additional questions that you hadn't thought of earlier. Afterward, be sure to send your interviewee a thank-you note.

THE HUNT IS ON

IN SOME CASES, CONDUCTING PRIMARY RESEARCH isn't as much a matter of creating research methods as it is a hunt for sources that already exist. This is true of books, articles, and artifacts, for example. But where do you begin to look for them?

For books, the obvious place to look is the library. If you are writing a paper about a specific author, for example, any work by that author could be considered a primary source. And an autobiography by anyone (even you!) is a primary source about that person. In addition, the letters, diaries, and sometimes photographs of many important people have been collected and published in book form, and these, too, are primary sources, as are the texts of original documents such as the Declaration of Independence.

Books are not the only primary sources that can be found in the library, though. **Contemporary** newspaper accounts of an event can also serve as primary sources, and most libraries keep back copies of at least a few national newspapers such as *The New York Times*, along with local newspapers. Rather than filling room after room with paper copies of these newspapers, many libraries preserve them on microfilm—a type of film that holds a tiny picture of the newspaper that can be viewed only with a special machine. The articles that appear in the paper are listed in a separate **index**—either in print or online. If you don't know where your library's microfilm collection is kept, ask your librarian.

Today, finding old newspaper articles is often even easier than tracking down microfilm copies. Many newspapers have put searchable copies of

their archives online. *The New York Times* archive, for example, links to newspaper articles dating back to 1851 (although some require a fee to view). The other option for viewing many articles online is to check your library's online **databases**. Many libraries subscribe to news databases that provide the full texts of articles from a number of newspapers—sometimes from around the world. In most cases, you can access these databases from any computer by simply logging in with your library card number.

If your topic falls into the category of public policy—such as laws or court rulings—your best primary sources might be government documents. Many libraries hold at least some government documents, especially those pertaining to local government, and you can find federal documents at specifically designated Federal Depository Libraries. These are usually university or public libraries that have been selected to house government publications on a variety of topics, from air quality reports to adoption laws.

NEWSPAPER ARTICLES

You can find a list of the depository libraries in your state on the Web site of the United States Government Printing Office (http://catalog.gpo.gov/fdlpdir/FDLPdir.jsp), where you can also search for government publications, some of which are available online. If you are looking for quantitative primary data about the U.S., the Census Bureau's site (www.census.gov) provides information on nearly every aspect of the American population, from income statistics to the number of households with pets.

In some cases, the information you need may not be in the form of words—or even numbers—at all. Perhaps you are looking for an image,

ORAL

sound recording, video, or artifact. An original work of art, for example, is a primary source for a paper about the work—or about the artist who created

HISTORIES

it. Photographs of an event or person are also primary sources. Even the images in an advertisement could serve as a primary source—if you were examining the portrayal of women as homemakers in 1950s magazine ads, for example. Oral histories—in which an interviewer

records a person's account of her life or an event she witnessed—can be invaluable when collecting information about someone who has died or whom you cannot interview yourself. In place of words, artifacts provide a glimpse of the objects people have created and used, from tools to weapons to decorative crafts, jewelry, and clothing. These items can reveal much about a person's social status or a society's culture and customs.

But where would you even begin to look for these types of non-print primary sources? Some libraries keep special collections of materials, often related to local history. These items may be housed in a separate room or area of the library, so you may need to check with a librarian in order to locate them. Even if your library does not have a special collection relevant to your topic, a **reference librarian** may be able to direct you to a collection held elsewhere, perhaps in a museum or an archive. Museums, of course, house large collections of paintings and other artworks, plus artifacts such as pottery, weapons, and tools.

Archives about a certain topic, time period, or person can be kept by a public institution, a private business, a family, or a government office. Some archives are open to the public; others are restricted. In the U.S., the National Archives and Records Administration operates several archives around the country, including in Washington, D.C.; Seattle, Washington; Atlanta, Georgia; and Chicago, Illinois (you can find more National Archives locations at www.archives.gov). If you happen to live near a National Archives facility (and are at least 14 years old, the minimum age to use the archives without special permission), you can plan a visit to conduct research into the nation's history and people, largely through public records. Such archives can

be especially useful for researching **genealogy**. If you are planning to visit an archive, it is important to do some research beforehand. If you can't tell the archivists what you are looking for, they will never find the information you need in the volumes upon volumes stored there!

For some archives, you can view catalogs of items online so that you'll know exactly what to ask for when you arrive. Even if you do not live near an archive, you can order copies of some information. And sometimes, the information you need may be closer than you think—in your family's archives. Does your grandfather still have his military uniform, for example, or does your mom have some of the clothes she

wore in high school? Such items can give you a taste of history—and of your grandfather's or mother's place in it!

To access the materials in some archives, you don't have to travel any farther than a computer. Although most archival material is accessible only by visiting a physical location, a growing number of archives are putting copies and photographs of their artifacts online. The National Archives Web site, for example, provides copies of documents such as the Constitution, as well as American Civil War photographs and other items, although its online collection represents only a fraction of the materials available at National Archives buildings. In addition, the American Memory site offered by the U.S. Library of Congress (memory.loc.gov/ammem/index.html) contains more than 9 million items—including books, photographs, maps, and sound and video recordings. The transcripts (or written copies) of many oral history projects can also be found online (try entering your search term followed by "oral history," such as "September 11 oral history"). In some cases, audio or video recordings of the interviews are available as well.

From new Web sites to old books and from experiments to interviews, primary research can provide you with the firsthand information you need to write about almost any topic. There is a world of information all around you. So venture out and ask questions, watch people, and dig up documents. Then, take what you've learned and put it into your writing. The result will be a work that proves old points, provides new insights, or raises additional questions to explore—in other words, it will be a work worth reading!

A TREASURE TROVE

If you are looking for primary documents about American history, chances are you will be able to find something on the Library of Congress's American Memory Web site (memory.loc.gov/ammem/index.html). You can search the site for specific topics or browse its holdings by time period, item type, or collection subject. Among its many artifacts from America's early history are a letter written around 1692 by men and women accused of witchcraft; letters from founding fathers such as John Adams, Thomas Jefferson, and James Madison; and copies of treaties made with American Indian tribes. Representing more recent history, the site also features television ads for Coca-Cola, oral histories by American officials who served overseas from the 1940s to the 1990s, and video and audio interviews with Americans following the September 11, 2001, terrorist attacks. Many of the collections also offer secondary sources, such as interpretive essays, to help put the artifacts in context.

GLOSSARY

anecdotes—short, often humorous, stories relating interesting events or incidents

archive—a collection of documents or items of historical importance, or the building (or online location) where such documents or items are stored

artifacts—objects made by humans (as opposed to naturally occurring objects)

autobiography—the story of a person's life, written by the person him- or herself

biased—having a preference for or dislike of a certain person or idea that prevents one from making impartial judgments of that person or idea

cite—to quote someone else's work as evidence for an idea or argument

contemporary—occurring or living during the same time period

controversial—causing or marked by disagreements and arguments

databases—organized collections of data, or information, stored on a computer

ethnicity—describing the cultural group to which a person belongs or from which he or she is descended

extrapolate—to apply known facts about an observed portion (or sample) of a group to come to conclusions about the entire group

field—outside of a workplace or other controlled environment, where a researcher can have direct contact with a subject being studied

genealogy—the study of family history

hypothesis—an idea or explanation that can be tested through further investigation

index—in the context of books and articles, a catalog or list of specific items, often arranged alphabetically and providing details about where to find them

leading—worded so as to "lead" someone to give a specific, desired response

loaded—containing a hidden, often emotionally charged, meaning designed to lead a person to answer in a specific way

reference librarian—a professional trained to find and organize information in a library and elsewhere

relevant—related or connected to the idea or topic being discussed

representative—consisting of a range of typical members of a group

variables—elements of an experiment that can change or vary

WEB SITES

American Memory from the Library of Congress

http://memory.loc.gov/ammem/index.html

Select a topic to perform general searches or practice using specific search terms to find materials.

National Archives: Research Our Records

http://www.archives.gov/research/

Explore the many tools and resources available from the National Archives.

Survey Methods

http://www.ischool.utexas.edu/~palmquis/courses/survey.html

Discover further information on how to conduct different types of surveys.

What Are Primary Sources?

http://www.yale.edu/collections_collaborative/primarysources/
primarysources.html

Find out more about different types of primary documents.

SELECTED BIBLIOGRAPHY

Anson, Chris M., Robert A. Schwegler, and Marcia F. Muth. *The Longman Writer's Bible: The Complete Guide to Writing, Research, and Grammar*. New York: Pearson Longman, 2006.

Ballenger, Bruce. *The Curious Researcher: A Guide to Writing Research Papers*. New York: Pearson Longman, 2004.

Booth, Wayne C., Gregory G. Colomb, and Joseph M. Williams. *The Craft of Research*. Chicago: University of Chicago Press, 2008.

Chernow, Barbara A. *Beyond the Internet: Successful Research Strategies*. Lanham, Md.: Bernan Press, 2007.

George, Mary W. *The Elements of Library Research: What Every Student Needs to Know*. Princeton, N.J.: Princeton University Press, 2008.

Lane, Nancy, Margaret Chisholm, and Carolyn Mateer. *Techniques for Student Research: A Comprehensive Guide to Using the Library*. New York: Neal-Schuman Publishers, 2000.

Rodrigues, Dawn, and Raymond J. Rodrigues. *The Research Paper: A Guide to Library and Internet Research*. Upper Saddle River, N.J.: Prentice Hall, 2003.

Toronto Public Library. *The Research Virtuoso: Brilliant Methods for Normal Brains*. Toronto: Annick Press, 2006.

INDEX